YOUR KNOWLEDGE

- We will publish your bachelor's and master's thesis, essays and papers

- Your own eBook and book - sold worldwide in all relevant shops

- Earn money with each sale

Upload your text at www.GRIN.com
and publish for free

Bibliographic information published by the German National Library:

The German National Library lists this publication in the National Bibliography; detailed bibliographic data are available on the Internet at http://dnb.dnb.de .

Imprint:

Copyright © 2010 GRIN Verlag
Print and binding: Books on Demand GmbH, Norderstedt Germany
ISBN: 9783668677326

This book at GRIN:

https://www.grin.com/document/154133

Steffen Rudigier

Corporate Social Responsibility. The case of Siemens and Gazprom

GRIN Verlag

GRIN - Your knowledge has value

Since its foundation in 1998, GRIN has specialized in publishing academic texts by students, college teachers and other academics as e-book and printed book. The website www.grin.com is an ideal platform for presenting term papers, final papers, scientific essays, dissertations and specialist books.

Visit us on the internet:

http://www.grin.com/

http://www.facebook.com/grincom

http://www.twitter.com/grin_com

Corporate Social Responsibility
The case of Siemens and Gazprom

Mapping the Global Environment

Steffen Rudigier
1/6/2010

Table of Contents

Executive Summary

Issues like climate change, human rights abuses and poverty dominate our media. Often blamed for responsibility for many of the world's urgent problems, companies are increasingly expected to address them. The term "Corporate Social Responsibility" has therefore become synonymous for initiatives towards sustainable development.

This report was established to examine how far corporate social responsibility and sustainable development of multinational corporations has been practiced in recent years, specifically the German Siemens Group and Russia's Gazprom which are the subjects of this investigation.

The paper first highlights some key issues of the stakeholder theory and stakeholder salience model, which is then applied to determine the companies' key stakeholders. The most important ones are analysed and compared in terms of power, legitimacy and urgency. Section two deals with the evaluation of the companies' responsible business approaches, pointing out the extent to which these have increased or decreased and the underlying reasons effecting change. The final section seeks to identify the substance of the described approaches by applying the Ponte et.al typology and finally analysing them using the two contrasting perspectives "Good for Business" and "Critical Development".

Findings in section one show that classifying and addressing stakeholders is fundamentally through managerial perception and the variable constitutional contexts in which the companies are enacted. The results also reveal that Siemens has to deal cautiously with a wider group of stakeholders than Gazprom, due to the different constitutional situations in Germany and Russia. Gazprom is most influenced by its main-stakeholder, the Russian Federation, whereas the power of other stakeholder groups is relatively low.

The investigation of Siemens's socially responsible business approaches in section two revealed that the increasing awareness of CSR in Germany and Siemens's corruption affair were the main drivers intensifying their recent business approaches. CSR in Russia is still immature; consequently Gazprom's initiatives are limited. However, CSR is expected to increase in Russia and Gazprom as a global player must pursue international standards.

Final section findings show Siemens's recent business approaches apparently foster sustainable development, since they largely align with company business operations. However - apart from several CSR awards – reliable data supporting that argument is elusive. In contrast, Gazprom's initiatives mostly fail to show substance, reinforced by its negative media image.

While CSR reporting and numerous CSR initiatives are undoubtedly "good for business", this paper also questioned the positive contribution of CSR towards sustainable development. Having analysed the "critical development" perspective, findings show these initiatives often lack substance due to poor planning and omission of consensual definition for CSR.

The future of CSR is hard to predict. The report shows that business initiatives can benefit society, but also reveals that they often fail to meet societal expectations. In a neo-liberal market system where shareholder values still override the world's urgent problems, improvements towards sustainable development are difficult to realise. There are no statutory regulations, only those codes of practice that the companies voluntarily sign up for. Thus, it needs the pressure of governments, the UN and NGOs to support reforms that help companies to better address future initiatives towards sustainable development.

Introduction

In recent years, climate change, human rights abuses and poverty have been widely discussed in our media. In this context, the term "corporate social responsibility" (CSR) has increasingly gained importance and global corporations have begun to intensify their CSR activities relating to areas such as environmental protection or human rights. (Kercher, 2007).

Among numerous definitions of CSR, the Business for Social Responsibility (Kottler & Lee, 2005, p. 3) argues that CSR can be seen as "operating a business in a manner that meets or exceeds the ethical, legal, commercial, and public expectations that society has of business." In the past CSR initiatives were often regarded as benevolence, simply achieved through writing a cheque and "doing good" as conveniently possible. However, there has been a change towards long-term commitments to specific social issues and initiatives supporting business goals and objectives, referring to a firm's relevant stakeholders (Kottler & Lee, 2005).

This report attempts to analyse the extent to which the two MNCs Siemens AG and OAO Gazprom carry out sustainable CSR to their stakeholders. Siemens, a global group in electronics and electrical engineering, is headquartered in Munich, Germany and employs about 427,000 people. The company operates in 190 countries, primarily in the industrial, energy and health care sectors (Datamonitor, 2009). Gazprom, headquartered in Moscow, Russia, is one the world's largest gas-producing companies with approximately 457,000 employees. The enterprise operates Russia's domestic gas pipeline network, delivering gas to countries across Central Asia and Europe, whereas Europe is the principal customer (Datamonitor, 2009).

Initially, the first section of this paper outlines the stakeholder salience theory and model, which is applied to exemplify Siemens and Gazprom stakeholders and their influence on the two companies. Section two is followed by a description of the companies' responsible business approaches, pointing out the extent to which these have increased or decreased and the underlying reasons effecting change. Section three seeks to identify the substance of the described approaches by applying the Ponte et.al typology and finally analysing them using the two contrasting perspectives "Good for Business" and "Critical Development".

1. Stakeholder Salience

Theory of stakeholder salience has gained broader acceptance and usage in the business environment and calls on companies to do business in a way that supports the needs and wants of certain stakeholders (O'Higgins & Morgan, 2006). In this context, Freeman (1984, p. 46) states "A stakeholder is any group or individual who can affect or is affected by the achievement of the organisation's objectives". Thereby, it is useful to classify stakeholder groups through evaluation of their represented interests, the power they possess relevant to the organisation or whether they hinder or support the achievement of the company's goals (Williams & Lewis, 2008). This process is called stakeholder mapping.

1.1 The Stakeholder Salience Model

A popular model identifying stakeholders and the degree paid to them by managers is the "The Stakeholder Salience Model" (Mitchell et al., 1997). This model classifies seven types[1] of stakeholder classes that emerge from various combinations of three attributes[2]:

- **Power**
- **Legitimacy**
- **Urgency**

Figure 1: Stakeholder Typology: One, Two, or Three Attributes Present

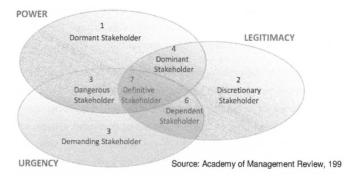

Source: Academy of Management Review, 199

---

[1]The lower salience classes (1, 2, 3) are defined as latent stakeholders, possessing only one of the three attributes and include dormant, discretionary and demanding stakeholders. The moderately classes (4, 5, 6) are defined as expectant stakeholders, showing two of the three attributes, and include dominant, dependent and dangerous stakeholders. The stakeholders possessing all of the three attributes (7) are called definitive stakeholders (Mitchell et.al, 1997).

[2] **Power** of the stakeholder relevant to the organisation; **Legitimacy** refers to socially accepted and expected structures or behaviours; **Urgency** or the degree to which stakeholder claims call for immediate attention (based on the stakeholder's time-sensitivity and criticality)

Mitchell et.al (1997) argue stakeholder classification is mainly based on managerial perception. However, the constitutional system in which the companies are enacted should also be considered (Amaeshi, 2007). So for instance, could the German coordinated market economy have a different influence on the power, legitimacy and urgency of stakeholders than the Russian post-communist economy, which is in transition to a market economy?

To discover who and what matters to companies, social or sustainability reports represent a prominent instrument highlighting the company's commitments in terms of sustainable business practices and their accountability to various stakeholders (Amaeshi, 2007). There have latterly been an increased number of such documents, providing responsible and transparent business practices to stakeholders. According to KPMG, a survey of the Global Fortune Top 250[3] companies illustrated that more than 80% now report on CSR (KPMG, 2008).

Stakeholder groups providing resources to the company essential for the companies' survival, are usually addressed more frequently compared to stakeholder groups which do not directly affect the daily business life (van Nimwegen et.al, 2008).

According to **Siemens's** sustainability report 2008, they address employees, customers, suppliers, investors, society, scientific community, government and NGOs as their key stakeholders (Siemens, 2008). Figure 2 provides a typology for stakeholder groups of Siemens in Germany. Further detailed discussion of some key stakeholder groups comparing the two companies will follow.

---

[3] The world's largest corporations, available at: http://money.cnn.com/magazines/fortune/global500/2009/full_list/

Figure 2: Stakeholder typology Siemens (Germany)

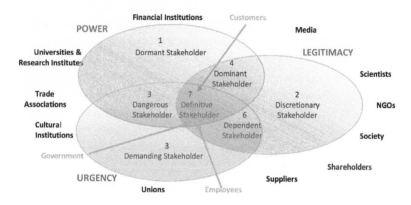

Siemens argues "our **employees** are our most important stakeholders" (Siemens, 2008). Relations between managers and employees are based on consensus, confidence and trust - perceived as a viable resource and not as a cost. Legitimacy therefore, is evaluated as high. Furthermore, in Germany, law permits workers' representatives in large companies seats on supervisory boards. These representatives have various abilities of influence with company management (Winkler, 2009). Therefore, employee power is high. Since workers' representatives can directly influence the management in terms of perceived problems, urgency too could be regarded as high, classifying employees as definitive stakeholders.

Siemens is embedded in a highly regulated business environment where **government** is regarded as a high power stakeholder having high legitimacy. According to Siemens, obeying law and legislation is its ultimate ambition (Siemens, 2008). Government claims require immediate attention, due to its high power and legitimacy thus, declaring it a definitive stakeholder (Winkler, 2009).

Siemens seeks to build long-lasting relationships with **customers** aligned with their needs and requirements (Siemens, 2008), which is resulting in a high legitimacy. Customers have consumer power since they can threaten the company's survival by boycotting products from Siemens. As a legitimated and to some extent powerful stakeholder, the customer could also be regarded as an urgent stakeholder, whose claims are dealt with very quickly. This qualifies customers as definitive stakeholders.

4

Gazprom, has lately been including CSR references in annual reports. Since 1996, the company also publishes annual environmental reports addressing stockholders, partners, public and social organisations as their stakeholders (Gazprom, 2008). However, compared to Siemens, the extent of addressing stakeholders is still relatively small and could be amended by additional groups, such as NGOs or suppliers. Figure 3 provides a typology of Gazprom's stakeholder groups.

Figure 3: Stakeholder typology Gazprom (Russia)

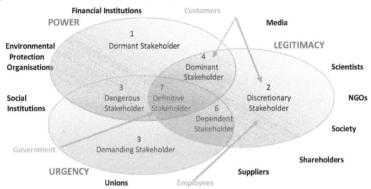

Gazprom regards its **employees** as respected and honoured stakeholders giving them legitimacy. Unlike Siemens, Gazprom employees have no right of direct co-determination or representation on the supervisory board (Burkhardt, 2007). Therefore, employee power is somewhat limited. Consequently, urgency might be regarded as relatively low, classifying employees as discretionary stakeholders.

Gazprom is a state-owned company. The Russian **government** controls the company with a majority stake of over 50% (Datamonitor, 2009). Unquestionably, this stakeholder possesses all three attributes, resulting in a definitive stakeholder.

Gazprom produces about 90% of Russia's gas with a monopoly on gas exports to Western and Eastern Europe (Country Monitor, 2005). **Customers** possess the attribute of legitimacy, but have little power or urgent claim, since Gazprom is the only gas distributor for Russian customers. However, this could change, since Gazprom considers raising prices, as a result of lacking liquidity, caused by the global recession. This might provoke the long-suffering Russian consumer, leading to social unrest, threatening the government and consequently Gazprom (Daly, 2009).

At first glance, stakeholder groups of both companies look similar. However, we can see that groups embedded in different constitutional contexts are perceived differently by managers.

2. Responsible Business Initiatives

Companies are increasingly expected to address urgent global problems like climate change, human rights abuses and poverty. As a result CSR has become a business approach for addressing the social and environmental impact of company activities (Freynas, 2009). Siemens and Gazprom are both reporting on business approaches which align with the interests of the broader society. These approaches will be highlighted below.

2.1 Responsible Business Initiatives of Siemens

Siemens refers to their United Nations Global Compact[4] membership, which reflects Siemens's commitment in socially responsible business practices. These practices are highlighted in table1.

Table 1: Siemens: Socially Responsible Business Activities

Category	Activities
Human Rights and anti-discrimination initiatives	• Business Conduct Guidelines to observe and control human rights and core work norms • Obligatory for all Siemens suppliers
Environmental Protection	• Medium-term plan (2006-2011) to improve carbon, energy and water performance by 20% • Raising employees' awareness of ecological responsibility inside and outside the company • Replacing dangerous substances with ecologically harmless substances
Environmentally friendly products	• Develop and market products and solutions enabling customers to reduce their CO_2 emissions
Anti-Corruption Measures	• Compliance Program for all employees (prevent, detect and respond)
Society Measures	• Education, volunteerism and disaster relief • Donations to defined humanitarian/social programs

Source: Siemens, Sustainability Report 2008

Table 1 illustrates that Siemens has been involved in a number of CSR practices recently. When the company first started to publish environmental reports in 1998, they were short and glossy, including only a limited amount of initiatives. In the last few years, in Germany, the awareness on CSR practices has been increasingly gaining importance (Pattberg, 2008), particularly in

---

[4] Launched in 2000 by the United Nations and Kofi Annan, the Global Compact serves as collaboration between corporations, UN agencies, NGOs, labour organisations and civil society to support universal environmental and social principles. These principles are represented by ten principles related to Human Rights, Labour Standards, Environment and Anti-Corruption (Banerjee, 2009)

sustainability, ecological and social issues (CSR Europe, 2009). Furthermore, NGOs in western countries have been aggressively putting pressure on MNCs, as a result of numerous corporate scandals, which have also alarmed a wider group of people (Littlechild, 2004). In 2006, Siemens itself was involved in Germany's biggest corruption scandal leading to costs of €1.3 billion and an immense loss in reputation (Gow, 2008). Subsequently, the company dramatically increased its CSR initiatives.

2.2 Responsible Business Initiatives of Gazprom

Gazprom's socially responsible business approaches are listed in table two below.

Table 2: Gazprom: Socially Responsible Business Activities

Category	Activities
Environmental Protection	• Developing effective technologies of energy resources saving • Complex Restoring of disturbed eco-systems at construction and exploitation sites • Cooperation with federal and regional legislative on improving of the environment protection legislation
Employees	• Complying with International Labour Conventions ratified by the Russian Federation • Professional education and training for employees
Society Measures	• Participation in social support projects • Support of economically disadvantageous people, servicemen, veterans and invalids • Investments in new production and social infrastructure facilities • Support of culture, sports, education and science • The "Gazprom to Children" program to back Russia's children and youth.

Source: Gazprom, Environmental and Annual Report 2008

In comparison, Gazprom is operating in a business environment where CSR practices have not yet assumed the significance of the West (Kuznetsov et.al, 2009). However, CSR is expected to gain future importance, since the Russian economy is global-oriented. Also, non-financial risks such as social conflicts or environmental sanctions are increasingly recognised as important factors for Russian companies (Kostin, 2007). Yet, only large companies – such as Gazprom - have realised that CSR and reporting on social issues could be strategically important to their businesses. Gazprom is one of the few Russian companies which have been providing separate environmental reports in recent years (Kostin, 2007). This might restore the company's reputation, especially in an industry often linked with environmental pollution. Furthermore, Gazprom is getting closer to the level of international standards even when reports are still underdeveloped, compared to those of western companies.

3. Analysis and Classification of Business Approaches

In their social reports, both, Siemens and Gazprom highlight carrying out responsibility to their stakeholders.   But, publishing a CSR report alone does not necessarily assume social responsibility (Banerjee, 2008).   Whether a company is actually conducting socially responsible business is dependent on the broader society perception and not just talking about it.

This section is using the Ponte et al. typology to identify the substance of the above-mentioned initiatives of both companies. This framework is divided into four different categories:

- Ranking in corporate governance/CSR awards
- Negative press in the past
- Engaged/Disengaged CSR profile for its initiatives
    - ➔ Engaged: the extent to which initiatives have direct impact on company operations
    - ➔ Disengaged: the extent to which initiatives are weakly linked to company operations such as fund-raising or corporate philanthropy
- Proximate/Distant CSR profile for its initiatives
    - ➔ Proximate: initiatives taking place within the corporation itself
    - ➔ Distant: initiatives taking place in communities where the company, its suppliers or stakeholders are not present

(Ponte et al., 2009)

Siemens's business initiatives

Siemens Activities	CSR awards/Rankings	Negative press	Engaged/Disengaged	Proximate/Distant
Human Rights	Ranked number 19 in the Covalence Ethical Ranking for the Industrial Goods + Services Sector, 3.Q.2009 (Covalence, 2009)	Allegation of selling high technology to the regime of Iran, which is using them to spy the public, violating human rights (Rhoads & Chao, 2009)	Engaged	Proximate
Environmental Protection	Highest scoring in the Carbon Disclosure Project (2009) of industrial companies. China Business News's "Excellent Enterprise for Environment Friendliness" Award, 2009 (Siemens, 2009)		Engaged	Proximate
Environmentally friendly products	German Sustainability Award for sustainable Development, 2008 (2009)		Engaged	Proximate
Anti-Corruption Measures		Involved in Germany's biggest corruption scandal. Bribes in the amount of €420 million (1999-2004) in Europe, Asia and Africa. (Welt, 2007). Violation against the 10th principle of the UN Global Compact	Engaged	Proximate
Society Measures	"China's Best Corporate Citizen" Award, 2009. (Siemens, 2009) China Charity Award, 2008 (Siemens, 2008) CSR Award for "Light for Life" Project in Pakistan, 2007 (Siemens, 2007)		Disengaged	Distant

Gazprom's business initiatives

Gazprom Activities	CSR awards/rankings	Negative press	Engaged/Diseng.	Proximate/Distant
Human Rights		Accused of developing gas fields in Iran, which is violating human rights and suspected to develop uranium for nuclear weapons (The Economist, 2009). Possession of the major media assets (TV/Radio channels, newspapers) →Censorship →Abuse of media for political propaganda (Meyer, 2007).	Disengaged	Distant
Environmental Protection		Plans to dig trenches for pipelines in Siberia, →could cause a fresh outbreak of anthrax →threat to life of Nenets nomads (Cosson, 2006). Plans to start extracting from the Yamal peninsula in north-western Siberia (Russia largest gas reserves) by 2011-2012 →WWF activists see serious environmental effects (Kohn, 2007).	Engaged	Proximate
Anti-corruption/blackmailing measures		In January 2009, Gazprom cut off Ukraine's gas, forcing them to sign a new energy deal with Moscow. Since EU countries are dependent on Russian gas piped through the Ukraine, this energy blackmail left European consumers in the cold and made them subject to Moscow's political pressure (Petersen, 2009). Involved in corruption (transfer pricing, asset stripping and kickbacks) (Aslund, 2008). →Corruption Perception Index 2009, Russia ranked 146 out of 180 (Transparency International, 2009)	Disengaged	Distant
Society Measures			Disengaged	Distant

The results show Siemens's self-regulated business approaches have been mostly "engaged" in recent years. By integrating more transparency, an anti-corruption programme and activities to reduce CO_2 emissions or developing environmental-friendly products with direct impact on sustainable development, the company endeavours to compensate a former lack of compliance to rules and regulations. Evidence is given by the awards listed above. However, without regular monitoring of labour and environmental standards, incidents like trading with Iran might be queried if a credible code of conduct is to produce any meaningful social outcomes.

Apart from a minority, Gazprom's activities largely fail to show substance. Reporting CSR initiatives seems more of a PR exercise. As a result, we can assume that the company's business approaches merely serve self-interest, obviously profit maximisation, based on a shareholder value basis.

3.1 "CSR is Good for Business" perspective

Theory states participation in corporate social initiatives can have positive effects on a company's reputation, its bottom line and the broader community (Kottler & Lee, 2005). Further evidence shows companies with responsible community attitude increasingly appeal to customers and employees, as much as investors or financial analysts (Rake, 2005). According to Porter & Kramer (2006) the right social initiatives along with a company's core strategies may achieve competitive advantage.

Siemens's former lack of transparency and social irresponsibility has dramatically impacted the company in terms of corruption costs, reputation loss, economic degradation, and misallocation of foreign investments. It believes business related social initiatives and transparent reporting form the cornerstone to regain public trust and improve profitability and growth in the long-term.

Also, Gazprom's largely philanthropic initiatives, such as funding money for social projects, sports or children could positively impact on the company's business.

3.2 "Critical Development" perspective

There exists a distorted view of CSR, in the assumption that initiatives always result in win-win situations for both companies and stakeholders (Prieto-Carrón et al., 2006). However, the relationship between CSR and development has been widely discussed in terms of potential and limitations of business initiatives towards social development (Idemudia, 2008). Often, initiatives lack substance, being unconnected with the company's business operations or not happening in communities where they and their stakeholders are embedded. Basically, inappropriate planning and the complexity and absence of consensual definitions of CSR are undermining long-term development objectives (Idemudia, 2008).

Evidence shows both Siemens and Gazprom investing in philanthropic or charitable projects, impacting the society at short-notice, but without substance for sustainable development.

5. Conclusion

Both companies operate in industries with high public visibility; CSR and sustainable development have increasing significance. Thereby, addressing key stakeholders is of utmost importance. The stakeholder salience model applied shows that classifying and addressing stakeholders is fundamentally through managerial perception and the variable constitutional contexts in which the companies are enacted.

The investigation of Siemens's socially responsible business approaches revealed that the increasing awareness of CSR in Germany and Siemens's corruption affair were the main drivers intensifying their recent business approaches. CSR in Russia is still immature; consequently Gazprom's initiatives are limited. However, CSR is expected to increase in Russia and Gazprom as a global player must pursue international standards.

Final section findings show Siemens's recent business approaches apparently foster sustainable development, since they largely align with company business operations. However - apart from several CSR awards –reliable data supporting that argument is elusive. In contrast, Gazprom's initiatives mostly fail to show substance, reinforced by its negative media image.

While CSR reporting and numerous CSR initiatives are undoubtedly "good for business", this paper also questioned the positive contribution of CSR towards sustainable development. Having analysed the "critical development" perspective, findings show these initiatives often lack substance due to poor planning and omission of consensual definition for CSR.

The future of CSR is hard to predict. The report shows that business initiatives can benefit society, but also reveals that they often fail to meet societal expectations. In a neo-liberal market system where shareholder values still override the world's urgent problems, improvements towards sustainable development are difficult to realise. There are no statutory regulations, only those codes of practice that the companies voluntarily sign up for. Thus, it needs the pressure of governments, the UN and NGOs to support reforms that help companies to better address future initiatives towards sustainable development.

Bibliography

Amaeshi, K., 2007. *Who Matters to UK and German Firms? Modelling Stakeholder Salience Through Corporate Social Reports* [Online] Available at: http://www2.warwick.ac.uk/fac/soc/csgr/research/workingpapers/2007/wp22707.pdf [Accessed 31 December 2009]

Aslund, A., 2008. Russia, Energy and the European Union: Perspectives on Gazprom. [Online] 15 May. Available at: http://www.iie.com/publications/papers/aslund0508.pdf [Accessed 4 January 2010].

Banerjee, S.B., 2009. *Corporate Social Responsibility; The Good The Bad and The Ugly.* Glos: Edward Elgar Publishing Limited.

Burkhardt, B., 2007. *Arbeitsrecht in Russland* [Online] Available at: http://www.bblaw.com/uploads/media/BB_LaborLaw_Russia_de.pdf [Accessed 2 January 2010].

CSR Europe, 2009. *A Guide to CSR in Europe; Country Insights by CSR Europe's National Partner Organisations.*[Online] Available at: http://www.csreurope.org/data/files/20091012_a_guide_to_csr_in_europe_final.pdf [Accessed 3 January 2010].

Carbon Disclosure Project, 2009. Global 500 Report [Online] Available at: https://www.cdproject.net/CDPResults/CDP_2009_Global_500_Report_with_Industry_Snapshots.pdf [Accessed 4 January 2010]

Cosson, 2006. Pipeline Threatens Nomads; Russia's Nenets Fear Exposure to Anthrax. *The Washington Times* [Online] 6 September. Available at: http://findarticles.com/p/articles/mi_hb5244/is_200609/ai_n19658782/ [Accessed 4 January 2010].

Daly, J., 2009. Analysis: Russia's rising gas prices. *Energy Daily* [Online] Available at: http://www.energy-daily.com/reports/Analysis_Russias_rising_energy_prices_999.html [Accessed 2 January 2010].

Datamonitor, 2009. *OAO Gazprom; Company Profile.* [Online] Available at: http://web.ebscohost.com.ezproxy.liv.ac.uk/ehost/pdf?vid=7&hid=103&sid=b06e11e9-5087-4b98-b087-f950052821c1%40sessionmgr112 [Accessed 30 December 2009].

Datamonitor, 2009. Siemens *AG; Company Profile.* [Online] Available at: http://web.ebscohost.com.ezproxy.liv.ac.uk/ehost/pdf?vid=8&hid=103&sid=b06e11e9-5087-4b98-b087-f950052821c1%40sessionmgr112 [Accessed 30 December 2009].

Deutscher Nachhaltigkeitspreis, 2009. [Online] Available at: http://www.deutscher-nachhaltigkeitspreis.de/index.php?cid=148&SID=687ca179b9a745c015cfda908278472b [Accessed 4 January 2010]

Freynas, J.G., 2009. Beyond Social Corporate Responsibility. *Oil Multinationals and Social Changes.* [Online] Available at: http://www.ewidgetsonline.com/dxreader/Reader.aspx?token=Wp8lutaZEJncx9Mnm2Gs0A%3d%3d&rand=803084683&buyNowLink=http%3a%2f%2fwww.cambridge.org%2fuk%2fcatalogue%2fAddToBasket.asp%3fisbn%3d9780521868440%26qty%3d1 [Accessed 3 January 2010].

Freeman, R.E., 1984. Strategic Management: A Stakeholder Approach. Boston: Pitman Publishing Ltd.

Gazprom, 2008. *Environment Protection Environmental Report 2008.* [Online] Available at: http://www.gazprom.com/f/posts/71/879403/1er_eng_2008.pdf [Accessed 2 January 2010].

Gazprom, 2008. *Annual Report 2008.* [Online] Available at: http://old.gazprom.ru/report2008/en/index.html#:/ref/ [Accessed 3 January 2010].

Gow, D., 2008. Scandal-hit Siemens now squeaky clean. *Guardian.co.uk.* [Online] 24 June. Available at: http://www.guardian.co.uk/business/2008/jun/24/siemens.bribery [Accessed 4 January 2010]

Idemudia, U., 2008. *Conceptualising the CSR and Development Debate.* [Online] Available at: http://media.web.britannica.com/ebsco/pdf/31/31954896.pdf [Accessed 5 January 2010].

Kercher, K., 2007. Corporate Social Responsibility: Impact of globalisation and international business. *Corporate Governance eJournal*. [Online] Available at: http://epublications.bond.edu.au/cgi/viewcontent.cgi?article=1003&context=cgej [Accessed 29 December 2009].

Kohn, M., 2007. The Arctic Killers. *New Statesman*. [Online] Available at: http://web.ebscohost.com.ezproxy.liv.ac.uk/ehost/detail?vid=4&hid=111&sid=97117cf3-7891-4ec8-8535-e391ccffdf96%40sessionmgr113&bdata=JnNpdGU9ZWhvc3QtbGl2ZQ%3d%3d#db=buh&AN=2 6150611 [Accessed 4 January 2010].

Kostin, A., 2007. Russia: The Evolving Corporate Social Responsibility Landscape. *The Global Compact*. [Online] Available at: http://www.enewsbuilder.net/globalcompact/e_article000775164.cfm [Accessed 3 January 2010].

KPMG International, 2008. *International Survey of Corporate Social Responsibility Reporting 2008*. [Online] Available at: http://www.kpmg.com/Global/en/IssuesAndInsights/ArticlesPublications/Documents/International -corporate-responsibility-survey-2008.pdf [Accessed 31 December 2009].

Kotler, P. Lee, N., 2005. *Corporate Social Responsibility; Doing the Most Good for Your Company and Your Cause*. Hoboken, New Jersey: John Wiley & Sons, Inc.

Kuznetsov, A. Kuznetsova, O., Warren, R., 2009. CSR and the legitimacy of business in transition economies: The case of Russia. *Scandivnavian Journal of Management* [Online] 25 pp. 37-45 Available at: http://www.sciencedirect.com/science?_ob=ArticleURL&_udi=B6VFS-4V995SC-1&_user=777686&_coverDate=03%2F31%2F2009&_rdoc=1&_fmt=full&_orig=search&_cdi=601 8&_sort=d&_docanchor=&view=c&_searchStrId=1151801840&_rerunOrigin=google&_acct=C00 0043031&_version=1&_urlVersion=0&_userid=777686&md5=9d60249648aaf3b5438064082dfa b569#bib13 [Accessed 2 January 2010].

Littlechild, M., 2004. The Reporting Dilemma. *EBF*. [Online] 16. Available at: http://web.ebscohost.com.ezproxy.liv.ac.uk/ehost/pdf?vid=4&hid=107&sid=8f9b5076-4f8d-4aba-8bcf-3a656c1ac2c7%40sessionmgr112 [Accessed 3 January 2010]

Meyer, H., 2007. Putin Tightens Internet Controls Before Presidential Election. *Bloomberg.com.* [Online] Available at:
http://www.bloomberg.com/apps/news?pid=20601100&sid=a2Zf7wMQnNQ4 [Accessed 4 January 2010].

Mitchell, R.K. Agle, B.R., Wood, D.J., 1997. Toward a Theory of Stakeholder Identification and Salience: Defining the Principle of Who and What really counts. *Academy of Management Review* [Online] 22 (4) pp. 853-886. Available at:
http://web.ebscohost.com.ezproxy.liv.ac.uk/ehost/pdf?vid=2&hid=107&sid=aa617328-366f-4260-89cd-61ea506bfc47%40sessionmgr113 [Accessed 30 December 2009].

O'Higgins, E.R.E. Morgan, J.W., 2006. Stakeholder salience and engagement in political organisations; Who and what really counts? *Society and Business Review* [Online] 1 (1) pp. 62-76. Available at:
http://www.emeraldinsight.com/Insight/viewPDF.jsp?contentType=Article&Filename=html/Output/Published/EmeraldFullTextArticle/Pdf/2960010105.pdf [Accessed 30 December 2009].

Pattberg, F., 2008. *Corporate Social Responsibility in Germany.* [Online] Available at:
http://www.fabianpattberg.com/2008/10/corporate-social-responsibility-csr-in-germany---part-1/ [Accessed 3 January 2010].

Petersen, A., 2009. Moscow's blackmail and blackout; without alternative fuel sources, we could be left out in the cold. *The Washington Times.* [Online] 31 July Available at:
http://www.washingtontimes.com/news/2009/jul/31/moscows-blackmail-and-blackout/ [Accessed 4 January 2010].

Ponte, S. Richey, L.A. Baab, M., 2009. Bono's (RED) Product Initiative: corporate social responsibility that solves the problems of "distant others". *Third World Quarterly.* [Online] (30) 2 pp. 301-317. Available at:
http://pdfserve.informaworld.com/989239_793400741_909179098.pdf [Accessed 4 January 2010].

Porter, M.E. Kramer, M.R., 2006. Strategy & Society; The Link Between Competitive Advantage and Corporate Social Responsibility. *Harvard Business Review.* [Online] Available at:
http://custom.hbsp.harvard.edu/custom_pdfs/FSGIMR0612D2006122113.pdf [Accessed 5 January 2010].

Prieto-Carròn, M., 2006. *Critical Perspectives on CSR and development: what we know and what we don't know.*[Online] Available at: http://www.scribd.com/doc/7291818/PrietoCarron-Critical-Perspectives-on-CSR-and-Development [Accessed 5 January 2010].

Rake, M., 2005. The business case for corporate social responsibility. *CSR Quest.* [Online] Available at: http://www.csrquest.net/default.aspx?articleID=10771&heading= [Accessed 5 January 2010].

Rhoads, C. Chao, L., 2009. Iran's Web Spying Aided by Western Technology. *The Wall Street Journal.* [Online] June 22. Available at: http://online.wsj.com/article/SB124562668777335653.html#mod=todays_us_page_one [Accessed 4 January 2010].

Siemens, 2007. *Siemens "Light for Life" Project Wins CSR Award.* [Online] Available at: http://www.siemens.com.pk/CSRLFL.html [Accessed 4 January 2010].

Siemens, 2008. *Sustainability Report 2008.* [Online] Available at: http://www.unglobalcompact.org/data/ungc_cops_resources/CD192523-4985-48A0-B458-5C6DEA30841F/COP.pdf [Accessed 31 December 2009].

Siemens China, 2008. [Online] Available at: http://cn.siemens.com/cms/cn/english/press/presscontent/Pages/20081211.aspx [Accessed 4 January 2010].

Siemens, 2009. *Siemens China honoured a number of prestigious CSR awards.* [Online] Available at: http://w1.siemens.com.cn/news_en/news_articles_en/1430.aspx [Accessed 4 January].

Transparency International, 2009. *Corruption Perceptions Index 2009.* [Online] Available at: http://www.transparency.org/policy_research/surveys_indices/cpi/2009/cpi_2009_table [Accessed 4 January 2010].

The Economist, 2009. Thank you, Mr Putin and Mr Hu; Diplomacy runs out with Iran. *Volume 393 Number 8660 p.12.*

Van Nimwegen, G. Bollen, L. Hassink, H. Thijssens, T., 2008. A Stakeholder Perspective on Mission Statements: an international emperical studay. *International Journal of Organizational Analysis.* [Online] Available at:
http://www.emeraldinsight.com/Insight/viewPDF.jsp?contentType=Article&Filename=html/Output/Published/EmeraldFullTextArticle/Pdf/3450160105.pdf [Accessed 1 January 2010].

Welt.online, 2007. *Was Siemens vorgeworfen wird.* [Online] 28 March. Available at:
http://www.welt.de/wirtschaft/article782332/Was_Siemens_vorgeworfen_wird.html [Accessed 5 January 2010].

Williams, W. Lewis, D., 2008. Strategic Management Tools and Public Sector Management; the challenge of context specificity. *Public Management Review.* [Online] 10 (5) pp. 653-671. Available at:
http://web.ebscohost.com.ezproxy.liv.ac.uk/ehost/pdf?vid=5&hid=107&sid=1f8ca4ca-3c23-46e4-b878-704f5c853d00%40sessionmgr110 [Accessed 30 December 2009]

CPSIA information can be obtained
at www.ICGtesting.com
Printed in the USA
LVHW031119240222
711913LV00007B/427